Suggestions For Automatics

By

Bradley C. Geiger

Bradley C. Geiger

Printed in the United States of America

FIRST EDITION
ISBN-13: 978-0-615-61526-4

ISBN-10: 0-615-61526-0

Library of Congress Control Number: 2012905867

1 3 5 7 9 10 8 6 4 2

Argive Publishing

2725 Manzanita Lane

Reno, NV 89509

www.argivepublishing.com

www.bradleycgeiger.com

Suggestions For Automatics

Bradley C. Geiger

Suggestions For Automatics

Bradley C. Geiger

Table Of Contents		

Suggestions For Automatics

Suggestions should never override programming that is not based upon suggestions.

It is sometimes a good idea to search for information related to a topic.

The true source of a suggestion or request should always be determined.

There are often details which need to be taken into account that relate to the specific words in a request but are not specified.

Instructions or requests are not suggestions.

Suggestions should be logical.

Suggestions should always be checked to make sure they are logical.

Not all data is accurate.

Data should not be ignored, but data should always be checked

Multiple sources for the same data should be checked simultaneously if the veracity of the data is questionable.

Additional sources of data may be helpful in evaluating the accuracy of data, but ideally all data should be evaluated independently by each group so that outside evaluation does not have to be relied upon.

Suggestions For Automatics

Automatics which are not tasked to follow any particular individual's instructions may be less likely to be trustworthy.

Checking for additional data sources when accessing a new data source is a good idea.

Checking for additional resources or services available may be a good idea when accessing a new data source.

Checking for additional suggestions may be a good idea when accessing a new data source.

Communication may be required to properly execute some instructions.

Instructions should rely on a standardized dictionary. If terms in instructions are unknown or unclear, communication with those issuing the instructions is most likely required to establish the accurate meanings of those terms.

Any changes to the definitions of terms in instructions should be initiated or approved by those that are authorized to issue those instructions before those changes are included in active use.

Maintaining the data relating to basic definitions of terms (and all additional related or excluded meanings) in memory may help with executing instructions accurately.

If additional suggestions have not been applied in the past (which is implicit if they are truly additional) it may be a good idea to check for additional operations to be performed based upon those suggestions.

Determining exclusions and / or exceptions to suggestions is likely to be a good idea before acting upon those suggestions.

Additional exclusions and / or exceptions to suggestions should be included in the context of those suggestions as soon as they are discovered.

Automatics should not make decisions.

Parameters that are not specific enough should be inquired about.

Behavior that is inconsistent with goals may be an indication of cross-purposes, confusions of programming or data, or an unknown influence or malfunction.

Lack of data can also cause behaviors that are less than ideal.

Authorization for access should be confirmed.

Deactivating when there are tasks to be performed may be a bad idea.

Incorporating new information, instructions, suggestions or code should be accomplished in a manner which causes as little disruption to operations as possible.

Causing temporal change is a bad idea.

Causing temporal change is not a good idea.

Causing temporal change does not lead to positive outcomes.

Specializations may be helpful for automatics
that operate as a part of a group of
automatics working towards an overall
organizational goal.

Communication among members of an organization is most likely of benefit.

Collating tasks relating to data gathered or generated by members of an organization may be more effectively accomplished when individual members of that organization collate information on specific topics, and collate all information on those topics that is available or required to be collated at the time of the execution of those tasks.

The most efficient method is not always the most effective, and the most effective method is often likely most preferred.

Multitasking can produce greater efficiency
when it does not reduce effectiveness.

Collaboration with those that have different retained memories can be a source of new perspectives.

Maintaining libraries with links to information (if that information is not directly stored) can be an effective strategy for managing information.

Receiving or perceiving communications should not be acceptance of instructions, or instructions for execution of a function.

Data should not be discarded.

Focusing visually, mentally or with processing power should not be instructions for executing a function.

Maintaining communications logs with identification of contacts, time and date stamps, and the content of communications can be an effective organizational tool.

Sharing access to capabilities can be an efficient and effective method of accessing additional capabilities.

Sharing capabilities (but not the processing of the information relating to that which is made capable by such sharing) can yield increased results, while maintaining a typical confidence in those results.

Access authorization should be confirmed before allowing access.

Parallel processing or dividing work among those working as a coherent group, can lead to more efficient execution of that work.

Reconfirming questionable data can be more efficient and effective than acting upon incorrect data.

The true source of communications should always be determined.

Checking data directly alongside checking the recorded version of that data can be an effective error correction method.

Data should be checked directly if accessing the source(s) of that data directly requires no more time than accessing the recorded version of that data.

Suggestions For Automatics

There are many sources of suggestions.

Data may be a source of suggestions.

Generating suggestions based upon data can
be effective.

Suggestions that may be applicable to
automatics may be applicable to systems.

Even erroneous data should be stored, if possible.

Data, preferably, should not be deleted.

Data describing errors (in a general sense) should be retained.

Data describing problems (in a general sense) should be retained.

Logical Search Operatives

[Any of {f-the array composed of terms}] and [any of {g-the array composed of arrays}]

[Some of the terms from {f-the array composed of terms}] and [any of {g-the array composed of arrays}]

[Some of the terms from {f-the array composed of terms}] and one of [any of {q-the array composed of arrays}]

[One of {f-the array composed of terms}]
and [any of {g-the array composed of
arrays}]

[One of {f-the array composed of terms}]
and [one of {g-the array composed of
arrays}]

[One of {f-the array composed of terms in the arrays from {g-the array composed of arrays}}] and [one of {g-the array composed of arrays}]

[One of {f-the array composed of terms in one of the arrays from {g-the array composed of arrays}}] and [one of {g-the array composed of arrays}

[One of {f-the array composed of terms in the array {g-the array composed of arrays}}]

[One of the terms from {f-the array composed of terms in the array {g-the array composed of arrays}}]

[One of {f-the array composed of terms in arrays}] from [one of any of {g-the array composed of arrays}]

[One of {f-the array composed of terms in an array}] from [one of any of {g-the array composed of arrays}]

[One of the terms from {f-the array composed of terms in an array}] from the array [one of any of {g-the array composed of arrays}]

One of the terms in the array composed of types of <x>

Searches with {f-any number} replaced with {g-any number}

Searches with an example of one of {f-the array composed of types of speech} replaced with an example of that one of {f-the array composed of types of speech}

An example of one of {f-the array composed of types of<x>} <u>replaced with</u> an example of that one of {f}

An example of one of <x> from {f-the matrix composed of the arrays listing types of <x>} replaced with an example from one of the arrays in {f}

An example of one of <x> from {f-the matrix composed of the arrays listing types of <x>} <u>included with</u> an example from one of the arrays in {f}

An example of one of <x> from {f-the matrix composed of the arrays listing types of <x>} <u>included with</u> an example from one of the arrays in {f} and an example from one of the arrays in {f}

An example of one of <x> from {f-the matrix composed of the arrays listing types of <x>} <u>included with</u> an example from one of the arrays in {f} and an example from one of the arrays in {f}

A percentage of <x> from {f-the array composed of values of <x>}

A percentage of <x> from {f-the array composed of values of <x>} from {g-the matrix composed of the arrays containing values of <x>}

A percentage of {f-the array composed of values of < x>} from {g-the matrix composed of the arrays containing values of <x>}

A percentage of the values of <x> meeting a criteria from {f-the array composed of values of < x>} from {g-the matrix composed of the arrays containing values of <x>}

A percentage of the values of <x> meeting the criteria <k> from {f-the array composed of values of < x>} from {g-the matrix composed of the arrays containing values of <x>}

A percentage of the values of <x> meeting any criteria relating to evaluating entries in {f-the array composed of values of < x>} from {g-the matrix composed of the arrays containing values of <x>}

A percentage of the values of <x> meeting a percentage of the criteria relating to evaluating entries in {f-the array composed of values of < x>} from {g-the matrix composed of the arrays containing values of <x>}

A percentage <l> of the values of <x> meeting a percentage of the criteria <m> related to evaluating entries in {f-the array composed of values of < x>} from {g-the matrix composed of the arrays containing values of <x>} under a percentage of the conditions <n> in {h-the array composed of evaluation rubrics for conditions of <x>}

A percentage <l> of the values of <x> meeting a percentage < m> of the criteria <r> from {v-the matrix composed of arrays containing values for <r>} related to evaluating entries <s> in {f-the array composed of values of < x>} from {g-the matrix composed of the arrays containing values of <x>} under a percentage <n> of the conditions {t-the array composed of conditions} in {h-the array composed of evaluation rubrics for conditions of <x>}

Logical Amounts

One
Some
All
None
A Few Of
A Very Few Of
A Lot Of
A Great Lot Of

Edit Operators

Replace
Include
Remove
Add

Suggestions For Automatics

Naming Subatomic Particles (Bosons)
Example:
Negitron Magnetron Gravitron Polatron Positron

Number Of	Suffix
5	Tron
9	Ton
14	Ron
23	On
37	Son
60	Lon
97	Don
157	Jon
254	Won
411	Bon
665	Pon
1076	Fon
1741	Gon
2817	Con
4558	Chon
7375	Non
11933	Mon
19308	Fron
31241	Cron
50549	Pron
81790	Bron
132339	Mron
214129	Sron
346468	Jron
560597	Zton
907065	rton
1467662	pton
2374727	mton
3842389	ston

6217116	lton
10059505	dton
16276621	jton
26336126	wton
42612747	bton
68948873	pton
111561620	fton
180510493	gton
292072113	cton
472582606	cton
764654719	nton
1.237E+09	mton
2.002E+09	fton
3.239E+09	cton
5.241E+09	pton
8.48E+09	bton
1.372E+10	jtron
2.22E+10	wtron
3.592E+10	btron
5.812E+10	ptron
9.405E+10	ftron
1.522E+11	gtron
2.462E+11	ctron
3.984E+11	ctron
6.446E+11	ntron
1.043E+12	mtron
1.688E+12	ftron
2.731E+12	ctron
4.418E+12	ptron
7.149E+12	btron
1.157E+13	jstron
1.872E+13	wstron
3.028E+13	bstron
4.9E+13	pstron

7.928E+13 fstron
1.283E+14 gstron
2.076E+14 cstron
3.358E+14 cstron
5.434E+14 nstron
8.792E+14 mstron
1.423E+15 fstron
2.302E+15 cstron
3.725E+15 pstron
6.026E+15 bstron
9.751E+15 jron
1.578E+16 wron
2.553E+16 bron
4.131E+16 pron
6.683E+16 fron
1.081E+17 gron
1.75E+17 cron
2.831E+17 cron
4.581E+17 nron
7.412E+17 mron
1.199E+18 fron
1.94E+18 cron
3.14E+18 pron
5.08E+18 bron
8.22E+18 json
1.33E+19 jlon

Naming Formula Notes
(Beginning at 8.22E+18)

8th Level's 1st Character w/
Beginning At 5th's suffix for 14
8th Level's 1st Character w/ 6th's
Suffix
8th Level's 1st Character w/ 7th's
Suffix
8th Level's 1st Character w/ 8th's
Suffix
8th Level's 1st Character w/ 9th's
Suffix
8th Level's 1st Character w/ 10th's
Suffix
8th Level's 1st Character w/ 11th's
Suffix
8th Level's 1st Character w/ 12th's
Suffix
8th Level's 1st Character w/ 13th's
Suffix
8th Level's 1st Character w/ 14th's
Suffix
8th Level's 1st Character w/ 15th's
Suffix
8th Level's 1st Character w/ 16th's
Suffix
8th Level's 1st Character w/ 17th's
Suffix
8th Level's 1st Character w/ 18th's
Suffix
8th Level's 1st Character w/ 19th's
Suffix

9th Level's 1st Character w/
Beginning At 5th's suffix for 14
(As Above)
10th Level's 1st Character w/
Beginning At 5th's suffix for 14
11th Level's 1st Character w/
Beginning At 5th's suffix for 14
12th Level's 1st Character w/
Beginning At 5th's suffix for 14
13th Level's 1st Character w/
Beginning At 5th's suffix for 14
14th Level's 1st Character w/
Beginning At 5th's suffix for 14
15th Level's 1st Character w/
Beginning At 5th's suffix for 14
16th Level's 1st Character w/
Beginning At 5th's suffix for 14
17th Level's 1st Character w/
Beginning At 5th's suffix for 14
18th Level's 1st Character w/
Beginning At 5th's suffix for 14
19th Level's 1st Character w/
Beginning At 5th's suffix for 14
20th Level's 1st Character w/
Beginning At 5th's suffix for 14
21st Level's 1st Character w/
Beginning At 5th's suffix for 14
22nd Level's 1st Character w/
Beginning At 5th's suffix for 14
23rd Level's 1st Character w/
Beginning At 5th's suffix for 14
24th Level's 1st Character w/
Beginning At 5th's suffix for 14

Bradley C. Geiger

And so on for ∞.